Preface

We are thrilled to present the "Science Olympiad All Chapters Combined MCQ Class 5", a carefully crafted resource to help students excel in Science Olympiad examinations. This book provides a comprehensive collection of Multiple-Choice Questions (MCQs) covering all chapters of the Class 5 science syllabus, ensuring students gain a strong understanding of key concepts. Science is about curiosity, exploration, and solving real-world problems.

This book aims to simplify learning and make it engaging, while also challenging students to think critically and apply their knowledge. The questions are tailored to the Olympiad format, preparing students for the structure and difficulty of competitive exams.

We hope this book serves as a reliable guide for Class 5 students, inspiring them to excel in Science Olympiad competitions and nurturing a lifelong love for science.

Table of Contents

Questions: Part 1

1. What is the process by which plants convert carbon dioxide and water into glucose and oxygen?

   a) Respiration
   b) Transpiration
   c) Photosynthesis
   d) Fermentation

2. Which of the following is a renewable energy source?
   a) Coal
   b) Oil
   c) Natural gas
   d) Solar energy

3. What is the chemical formula for table salt?

   a) NaCl
   b) KCl
   c) NaOH
   d) HCl

4. What type of animal is a spider?

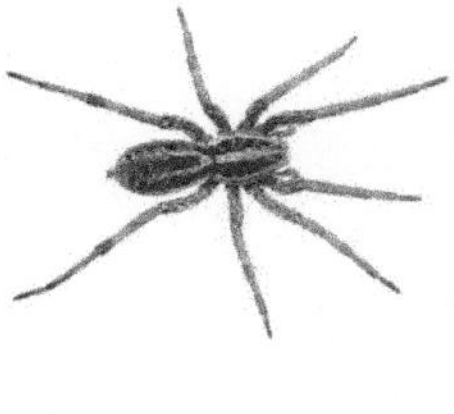

a) Insect

b) Arachnid

c) Mammal

d) Bird

5. What is the largest planet in our solar system?

a) Earth

b) Saturn

c) Jupiter

d) Neptune

6. What is the process by which water changes from a liquid to a gas?

a) Condensation

b) Freezing

c) Evaporation

d) Precipitation

7. Which organ in the human body is responsible for pumping blood?

a) Brain

b) Heart

c) Liver

d) Kidney

8. What type of rock is formed from the cooling of molten lava or magma?

a) Sedimentary

b) Metamorphic

c) Igneous

d) Fossil

9. What is the primary gas found in the Earth's atmosphere?

a) Oxygen

b) Carbon dioxide

c) Nitrogen

d) Hydrogen

10. What do we call the study of plants?

a) Zoology

b) Botany

c) Geology

d) Meteorology

11. Which planet is known as the Earth's twin?

a) Mars

b) Venus

c) Mercury

d) Jupiter

12. What is the main function of the roots of a plant?

a) Make food

b) Absorb water and nutrients

c) Support the plant

d) Produce seeds

13. What is the hardest natural substance on Earth?

a) Gold

b) Iron

c) Diamond

d) Silver

14. What type of energy is stored in food?

a) Kinetic energy

b) Potential energy

c) Chemical energy

d) Electrical energy

15. Which part of the plant is responsible for photosynthesis?

a) Roots

b) Stem

c) Leaves

d) Flowers

16. What do we call the force that pulls objects toward the center of the Earth?

a) Magnetism

b) Gravity

c) Friction

d) Inertia

17. Which planet is known for its rings?

a) Mars

b) Venus

c) Saturn

d) Jupiter

18. What is the main source of energy for the Earth?

a) The Moon

b) The Sun

c) The Stars

d) The Wind

19. What is the process by which plants make their food?

a) Respiration

b) Photosynthesis

c) Germination

d) Pollination

20. What type of animal is a frog?

a) Mammal

b) Reptile

c) Amphibian

d) Bird

21. What is the term for the variety of different types of life found on Earth?

a) Ecosystem

b) Habitat

c) Biodiversity

d) Population

22. What is the boiling point of water in degrees Celsius?

a) 50°C

b) 75°C

c) 100°C

d) 125°C

23. What is the main function of the human respiratory system?

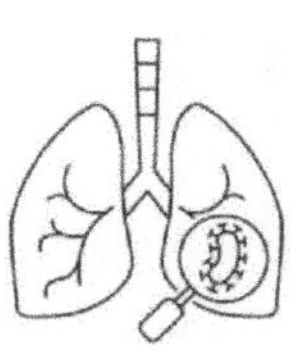

a) Digest food

b) Circulate blood

c) Exchange gases

d) Filter waste

24. What is the primary component of the human skeletal system?

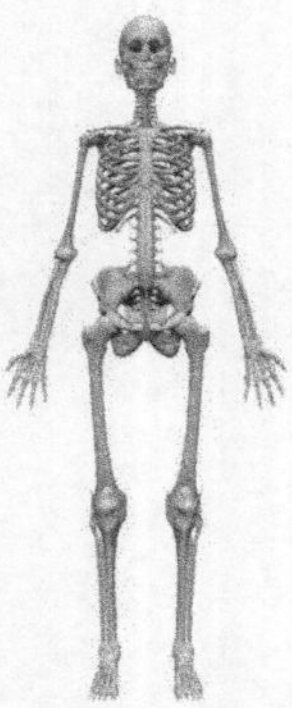

a) Muscle

b) Bone

c) Cartilage

d) Tendon

25. Which planet is closest to the Sun?

a) Earth

b) Venus

c) Mars

d) Mercury

26. What type of animal is a bat?

a) Bird

b) Mammal

c) Reptile

d) Amphibian

27. What is the name of the process by which cells divide to form new cells?

a) Photosynthesis

b) Respiration

c) Mitosis

d) Digestion

28. What is the term for the layer of gases that surrounds the Earth?

a) Hydrosphere

b) Lithosphere

c) Atmosphere

d) Biosphere

29. What type of energy comes from the Sun?

a) Kinetic energy

b) Solar energy

c) Electrical energy

d) Chemical energy

30. Which organ in the human body filters waste from the blood?

a) Heart

b) Liver

c) Lungs

d) Kidneys

31. What do we call the process by which water vapor turns into liquid water?

a) Evaporation

b) Condensation

c) Precipitation

d) Transpiration

32. What is the largest organ in the human body?

a) Heart

b) Liver

c) Skin

d) Brain

33. Which planet is known as the Red Planet?

a) Earth

b) Mars

c) Jupiter

d) Venus

34. What is the main function of the human digestive system?

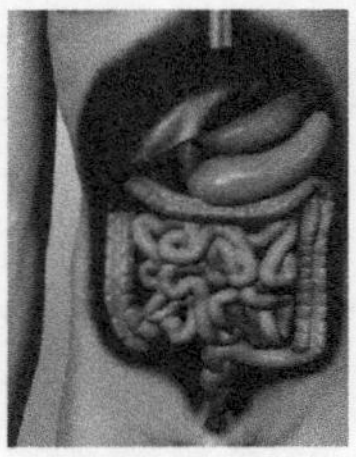

a) Pump blood

b) Exchange gases

c) Break down food

d) Filter waste

35. What type of animal is a whale?

a) Fish

b) Bird

c) Mammal

d) Reptile

36. What is the main component of the Earth's core?

a) Iron

b) Nickel

c) Oxygen

d) Silicon

37. What is the chemical formula for carbon dioxide?

a) CO

b) CO_2

c) C_2O

d) CO_3

38. What do we call a scientist who studies the weather?

a) Geologist

b) Meteorologist

c) Biologist

d) Chemist

39. Which part of the plant anchors it to the ground?

a) Leaf

b) Stem

c) Root

d) Flower

40. What type of rock is formed from layers of sediment?

a) Igneous

b) Metamorphic

c) Sedimentary

d) Fossil

41. What is the term for animals that eat only plants?

a) Herbivores

b) Carnivores

c) Omnivores

d) Insectivores

42. Which gas do plants use during photosynthesis?

a) Oxygen

b) Nitrogen

c) Carbon dioxide

d) Hydrogen

43. What is the term for a baby cow?

a) Kid

b) Calf

c) Foal

d) Lamb

44. What is the term for a scientist who studies animals?

a) Zoologist

b) Botanist

c) Geologist

d) Chemist

45. Which part of the human body is used for thinking?

a) Heart

b) Brain

c) Stomach

d) Liver

46. What is the primary function of leaves on a plant?

a) Absorb water

b) Support the plant

c) Make food

d) Produce seeds

47. What is the main gas found in the air we breathe?

a) Oxygen

b) Nitrogen

c) Carbon dioxide

d) Hydrogen

48. What is the process of breaking down food in the body called?

a) Respiration

b) Digestion

c) Circulation

d) Excretion

49. Which planet has the most moons?

a) Earth

b) Mars

c) Jupiter

d) Saturn

50. What type of animal is a lizard?

a) Mammal

b) Bird

c) Reptile

d) Fish

Questions: Part 2

51. What do we call the force that resists the motion of objects?

    a) Gravity

    b) Magnetism

    c) Friction

    d) Inertia

52. What is the chemical symbol for gold?

    a) Au

    b) Ag

    c) Fe

    d) Pb

53. Which part of the plant is responsible for reproduction?

    a) Leaf

    b) Stem

    c) Root

    d) Flower

54. What do you call a scientist who studies the stars and planets?

    a) Geologist

    b) Biologist

    c) Astronomer

    d) Meteorologist

55. Which planet is known as the Blue Planet?

    a) Mars

    b) Earth

    c) Neptune

    d) Uranus

56. What is the name of the process by which plants lose water vapor through small openings in their leaves?

    a) Photosynthesis

    b) Respiration

    c) Transpiration

d) Absorption

57. What is the main source of energy for the water cycle?

a) The Moon

b) The Sun

c) The Wind

d) The Earth's core

58. Which planet is the largest in our solar system?

a) Earth

b) Saturn

c) Jupiter

d) Neptune

59. What type of animal is a dolphin?

a) Fish

b) Mammal

c) Reptile

d) Amphibian

60. What is the main function of the human heart?

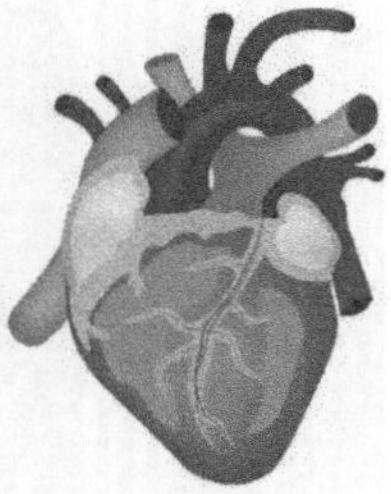

a) Digest food

b) Pump blood

c) Filter waste

d) Control movement

61. What type of rock is granite?

a) Igneous

b) Sedimentary

c) Metamorphic

d) Fossil

62. What is the term for the layer of the Earth below the crust?

a) Core

b) Mantle

c) Lithosphere

d) Biosphere

63. What is the term for the movement of water from the ground to the sky and back again?

a) Carbon cycle

b) Water cycle

c) Nitrogen cycle

d) Oxygen cycle

64. What is the process by which plants make their food using sunlight?

a) Respiration

b) Photosynthesis

c) Transpiration

d) Digestion

65. Which part of the plant contains chlorophyll?

a) Roots

b) Stem

c) Leaves

d) Flower

66. What is the chemical formula for water?

a) H_2O

b) CO_2

c) O_2

d) N_2

67. What do we call the group of bones that protect the brain?

a) Rib cage

b) Spine

c) Skull

d) Pelvis

68. What is the smallest unit of life?

a) Atom

b) Molecule

c) Cell

d) Organ

69. What type of energy is produced by moving water?

a) Solar energy

b) Wind energy

c) Hydroelectric energy

d) Geothermal energy

70. What is the term for the part of the Earth where life exists?

a) Atmosphere

b) Lithosphere

c) Hydrosphere

d) Biosphere

71. What is the name of our galaxy?

a) Andromeda

b) Milky Way

c) Triangulum

d) Whirlpool

72. What is the process of changing from a solid to a liquid?

a) Freezing

b) Evaporation

c) Melting

d) Condensation

73. What is the main function of the human liver?

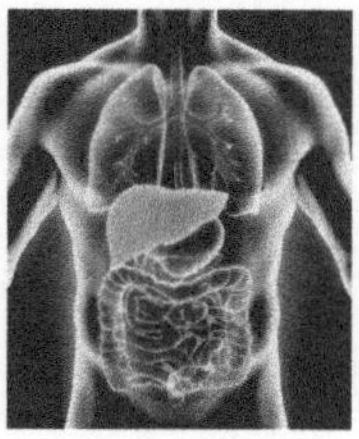

a) Pump blood

b) Filter waste

c) Exchange gases

d) Control movement

74. What do we call the path of an object as it moves around another object in space?

a) Rotation

b) Orbit

c) Revolution

d) Axis

75. What is the chemical formula for carbon dioxide?

a) CO

b) CO_2

c) C_2O

d) CO_3

76. What type of animal is a chameleon?

a) Mammal

b) Bird

c) Reptile

d) Fish

77. What is the term for the energy an object has due to its motion?

a) Potential energy

b) Kinetic energy

c) Chemical energy

d) Thermal energy

78. What do we call a baby sheep?

a) Calf

b) Foal

c) Lamb

d) Kid

79. What is the main component of the Earth's atmosphere?

a) Oxygen

b) Nitrogen

c) Carbon dioxide

d) Argon

80. Which planet is known as the Earth's twin?

a) Mars

b) Venus

c) Mercury

d) Jupiter

81. What is the term for the breaking down of rocks into smaller pieces?

a) Erosion

b) Weathering

c) Sedimentation

d) Fossilization

82. What is the primary function of red blood cells?

a) Fight infection

b) Carry oxygen

c) Clot blood

d) Provide nutrients

83. What type of energy is stored in a battery?

a) Solar energy

b) Electrical energy

c) Chemical energy

d) Kinetic energy

84. What do we call a scientist who studies living organisms?

a) Geologist

b) Biologist

c) Physicist

d) Chemist

85. What is the primary gas found in the air we exhale?

a) Oxygen

b) Nitrogen

c) Carbon dioxide

d) Hydrogen

86. What is the smallest bone in the human body?

a) Femur

b) Rib

c) Stapes

d) Skull

87. What is the process by which plants release oxygen into the air?

a) Respiration

b) Transpiration

c) Photosynthesis

d) Germination

88. What is the main function of white blood cells?

a) Carry oxygen

b) Fight infection

c) Clot blood

d) Provide nutrients

89. What type of animal is a kangaroo?

a) Reptile

b) Mammal

c) Amphibian

d) Bird

90. Which planet is known for having a prominent ring system?

a) Jupiter

b) Uranus

c) Saturn

d) Neptune

91. What is the primary gas found in the Earth's atmosphere?

a) Oxygen

b) Nitrogen

c) Carbon dioxide

d) Hydrogen

92. What is the boiling point of water?

a) 0°C

b) 50°C

c) 100°C

d) 150°C

93. What is the name of the process by which a caterpillar changes into a butterfly?

a) Photosynthesis

b) Metamorphosis

c) Respiration

d) Transpiration

94. What type of animal is a penguin?

a) Mammal

b) Bird

c) Fish

d) Reptile

95. What do we call the smallest part of an element?

a) Atom

b) Molecule

c) Cell

d) Organ

96. What is the main function of the human digestive system?

a) Pump blood

b) Break down food

c) Exchange gases

d) Filter waste

97. What type of energy is produced by the Sun?

a) Wind energy

b) Solar energy

c) Chemical energy

d) Nuclear energy

98. What do we call the part of the plant that contains seeds?

a) Stem

b) Root

c) Flower

d) Leaf

99. What is the main function of the human respiratory system?

a) Digest food

b) Circulate blood

c) Exchange gases

d) Filter waste

100. What is the term for animals that eat both plants and animals?

a) Herbivores

b) Carnivores

c) Omnivores

d) Insectivores

Questions – Part 3

101. Which gas do humans breathe in for survival?

a) Carbon dioxide

b) Oxygen

c) Nitrogen

d) Hydrogen

102. Which planet is known as the Red Planet?

a) Venus

b) Jupiter

c) Mars

d) Mercury

103. Which part of the plant makes food?

a) Stem

b) Roots

c) Leaves

d) Flowers

104. What is the process of water changing into vapor called?

a) Condensation

b) Evaporation

c) Precipitation

d) Filtration

105. Which is the largest organ in the human body?

a) Heart

b) Skin

c) Brain

d) Liver

106. What do bees collect from flowers to make honey?

a) Pollen

b) Nectar

c) Water

d) Seeds

107. Which force pulls objects toward the Earth?

a) Friction

b) Gravity

c) Magnetism

d) Electricity

108. Which is the main source of energy for Earth?

a) Moon

b) Sun

c) Stars

d) Wind

109. What do we call animals that eat both plants and animals?

a) Herbivores

b) Carnivores

c) Omnivores

d) Decomposers

110. Which is the fastest land animal?

a) Cheetah

b) Lion

c) Tiger

d) Horse

111. Which planet is closest to the Sun?

a) Earth

b) Mercury

c) Mars

d) Venus

112. Which blood cells help fight infections?

a) Red blood cells

b) White blood cells

c) Platelets

d) Plasma

113. Which type of star is the Sun?

a) Red giant

b) Yellow dwarf

c) Neutron star

d) Supernova

114. Which part of the human body helps in breathing?

a) Stomach

b) Lungs

c) Liver

d) Kidneys

115. What is the boiling point of water?

a) 50°C

b) 80°C

c) 100°C

d) 120°C

116. Which is the hardest natural substance on Earth?

a) Gold

b) Diamond

c) Iron

d) Silver

117. What is the study of stars and planets called?

a) Biology

b) Astronomy

c) Geology

d) Chemistry

118. Which gas do plants release during photosynthesis?

a) Carbon dioxide

b) Oxygen

c) Nitrogen

d) Hydrogen

119. Which part of the Earth is made up of molten rock?

a) Crust

b) Mantle

c) Core

d) Surface

120. Which organ pumps blood through the body?

a) Brain

b) Heart

c) Lungs

d) Liver

121. Which is the largest planet in the solar system?

a) Earth

b) Saturn

c) Jupiter

d) Neptune

122. Which part of the plant absorbs water and minerals?

a) Stem

b) Roots

c) Flowers

d) Leaves

123. Which is the smallest bone in the human body?

a) Femur

b) Stapes

c) Ulna

d) Radius

124. Which gas is used by fire to keep burning?

a) Carbon dioxide

b) Oxygen

c) Nitrogen

d) Helium

125. Which energy do we get from moving air?

a) Solar energy

b) Wind energy

c) Hydro energy

d) Heat energy

126. Which planet is known as the "Blue Planet"?

a) Earth

b) Neptune

c) Uranus

d) Venus

127. Which is the main organ of the circulatory system?

a) Brain

b) Lungs

c) Heart

d) Kidney

128. Which organ is called the control center of the body?

a) Heart

b) Brain

c) Liver

d) Lungs

129. Which force slows down moving objects?

a) Gravity

b) Friction

c) Magnetism

d) Air pressure

130. Which is the outermost layer of the Earth?

a) Mantle

b) Crust

c) Core

d) Surface

131. Which is the largest mammal on Earth?

a) Elephant

b) Blue whale

c) Hippopotamus

d) Giraffe

132. Which type of rock is formed from cooled lava?

a) Igneous rock

b) Sedimentary rock

c) Metamorphic rock

d) Fossil rock

133. Which star is closest to Earth?

a) Moon

b) Sun

c) Sirius

d) Polaris

134. Which gas is known as the "lifeline of Earth"?

a) Oxygen

b) Carbon dioxide

c) Nitrogen

d) Hydrogen

135. Which organ helps in filtering blood?

a) Lungs

b) Kidneys

c) Liver

d) Heart

136. Which planet is called the "Morning Star"?

a) Mars

b) Venus

c) Mercury

d) Jupiter

137. Which is the fastest bird in the world?

a) Eagle

b) Peregrine falcon

c) Owl

d) Hawk

138. Which organ helps in digestion of food?

a) Heart

b) Stomach

c) Brain

d) Kidney

139. Which is the smallest planet in the solar system?

a) Mars

b) Mercury

c) Pluto

d) Venus

140. Which gas do humans exhale while breathing?

a) Oxygen

b) Carbon dioxide

c) Nitrogen

d) Helium

141. Which part of the plant is called the food factory?

a) Stem

b) Roots

c) Leaves

d) Flowers

142. Which instrument is used to measure temperature?

a) Barometer

b) Thermometer

c) Rain gauge

d) Speedometer

143. Which organ in the human body helps us smell?

a) Eyes

b) Nose

c) Tongue

d) Skin

144. Which is the main source of light for the Earth?

a) Moon

b) Sun

c) Stars

d) Fire

145. Which force keeps planets in orbit around the Sun?

a) Friction

b) Gravity

c) Magnetism

d) Air pressure

146. Which is the largest bird in the world?

a) Ostrich

b) Eagle

c) Peacock

d) Swan

147. Which process changes water vapor into liquid water?

a) Evaporation

b) Condensation

c) Freezing

d) Melting

148. Which energy do we get from the Sun?

a) Solar energy

b) Wind energy

c) Hydro energy

d) Nuclear energy

149. Which organ helps us hear sounds?

a) Ear

b) Eye

c) Nose

d) Tongue

150. Which type of energy is stored in food?

a) Heat energy

b) Chemical energy

c) Kinetic energy

d) Potential energy

Answers

1. c) Photosynthesis
2. d) Solar energy
3. a) NaCl
4. b) Arachnid
5. c) Jupiter
6. c) Evaporation
7. b) Heart
8. c) Igneous
9. c) Nitrogen
10. b) Botany
11. b) Venus
12. b) Absorb water and nutrients
13. c) Diamond
14. c) Chemical energy
15. c) Leaves
16. b) Gravity
17. c) Saturn
18. b) The Sun
19. b) Photosynthesis
20. c) Amphibian
21. c) Biodiversity
22. c) 100°C
23. c) Exchange gases
24. b) Bone
25. d) Mercury
26. b) Mammal
27. c) Mitosis
28. c) Atmosphere
29. b) Solar energy
30. d) Kidneys

31. b) Condensation
32. c) Skin
33. b) Mars
34. c) Break down food
35. c) Mammal
36. a) Iron
37. b) CO_2
38. b) Meteorologist
39. c) Root
40. c) Sedimentary
41. a) Herbivores
42. c) Carbon dioxide
43. b) Calf
44. a) Zoologist
45. b) Brain
46. c) Make food
47. b) Nitrogen
48. b) Digestion
49. c) Jupiter
50. c) Reptile
51. c) Friction
52. a) Au
53. d) Flower
54. c) Astronomer
55. b) Earth
56. c) Transpiration
57. b) The Sun
58. c) Jupiter
59. b) Mammal
60. b) Pump blood
61. a) Igneous
62. b) Mantle

63. b) Water cycle
64. b) Photosynthesis
65. c) Leaves
66. a) H_2O
67. c) Skull
68. c) Cell
69. c) Hydroelectric energy
70. d) Biosphere
71. b) Milky Way
72. c) Melting
73. b) Filter waste
74. b) Orbit
75. b) CO_2
76. c) Reptile
77. b) Kinetic energy
78. c) Lamb
79. b) Nitrogen
80. b) Venus
81. b) Weathering
82. b) Carry oxygen
83. c) Chemical energy
84. b) Biologist
85. c) Carbon dioxide
86. c) Stapes
87. c) Photosynthesis
88. b) Fight infection
89. b) Mammal
90. c) Saturn
91. b) Nitrogen
92. c) 100°C
93. b) Metamorphosis
94. b) Bird

95. a) Atom
96. b) Break down food
97. b) Solar energy
98. c) Flower
99. c) Exchange gases
100. c) Omnivores
101. b) Oxygen
102. c) Mars
103. c) Leaves
104. b) Evaporation
105. b) Skin
106. b) Nectar
107. b) Gravity
108. b) Sun
109. c) Omnivores
110. a) Cheetah
111. b) Mercury
112. b) White blood cells
113. b) Yellow dwarf
114. b) Lungs
115. c) 100°C
116. b) Diamond
117. b) Astronomy
118. b) Oxygen
119. b) Mantle
120. b) Heart
121. c) Jupiter
122. b) Roots
123. b) Stapes
124. b) Oxygen
125. b) Wind energy
126. a) Earth

127. c) Heart
128. b) Brain
129. b) Friction
130. b) Crust
131. b) Blue whale
132. a) Igneous rock
133. b) Sun
134. a) Oxygen
135. b) Kidneys
136. b) Venus
137. b) Peregrine falcon
138. b) Stomach
139. b) Mercury
140. b) Carbon dioxide
141. c) Leaves
142. b) Thermometer
143. b) Nose
144. b) Sun
145. b) Gravity
146. a) Ostrich
147. b) Condensation
148. a) Solar energy
149. a) Ear
150. b) Chemical energy

9 789369 725373

Printed at Repro India Ltd.